All wood flooring does the same thing—it gives you a place to put your feet. So the choice between the three types—solid, engineered, and parquet flooring—comes down to how much you want to pay, what look you want, and where you're installing it.

The illustrations here show the basics of floor structure.

You'll find information about the different types of wood flooring on pages 2 to 6, and about the tools and other stuff you'll need on pages 7 to 11. Step-by-step installation details start on page 12.

Write notes or questions on the pages as you go along, and keep track of stuff you need on the back cover.

A GLUE-DOWN FLOOR

The standard installation method for parquet floors and some engineered strip or plank flooring. Adhesive is spread on the subfloor for tiles or individual pieces.

tiles

Adhesive

Joists

Plywood subfloor

A FLOATING FLOOR

The preferred installation method for some engineered strip and plank flooring. Boards are glued to each other, not fastened to the subfloor. They're held in place by their own weight.

Engineered wood boards

Joists

Foam underlayment

Existing flooring

Plywood subfloor

A NAILED STRIP OR PLANK FLOOR

The standard installation method for solid wood floors and an optional method for some engineered flooring. Strips and planks are blind-nailed through the tongue into the subfloor. Planks wider than about 4 inches are both blind-nailed and screwed through the face.

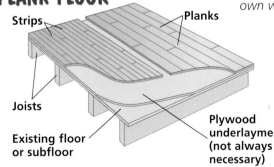

Strips

Planks

Joists

Existing floor or subfloor

Plywood underlayment (not always necessary)

1

SOLID WOOD FLOORING

Solid wood flooring comes in a range of widths—narrow boards are called strips and wider ones are planks. Both are nailed or stapled through the tongue into the subfloor. Planks wider than about 4 inches must be nailed or screwed through the face to prevent warping. Wood plugs fill the holes.

Boards come unfinished or prefinished. With unfinished boards, you get a wider range of sizes, species, and finish options. The downside? You have to sand and finish them, which is messy and time-consuming. With prefinished boards, you have fewer choices, but you get a floor you can walk on right away.

WHAT'S ON THE SIDE?

Most wood floor boards have a tongue on one side and one end, and a groove on the other side and end. Some people call the boards tongue and groove, and others call them "side and end matched" because they fit together both at the sides and at the ends.

Prefinished boards may also come with beveled edges. Bevels, sometimes called microbevels, help camouflage any minor differences in height from one board to the next.

Side and end matched

Beveled edge

STRIPS

Strip flooring comes in widths from 2$\frac{1}{4}$ inches to 3 inches and in a variety of stains and finishes. The most common size is $\frac{3}{4}$ inch thick and 2$\frac{1}{4}$ inches wide, but most home centers offer a selection. For greater variety in species, width, or finish, try a flooring dealer.

PLANKS

Plank flooring starts at about 3 inches wide. Boards of different widths are often used together in a plank floor. Planks may be more difficult to find and more expensive than strips.

ENGINEERED WOOD FLOORING

Engineered wood flooring looks like solid wood from the top, but it's made up of several layers. The top layer is hardwood; the other wood layers are less attractive and less expensive, but are also very stable. The layers are glued together with their grain running in opposite directions, which reduces the amount of expansion and contraction.

The flooring is cut tongue and groove, and comes in fixed or random lengths. It can be nailed or stapled down *(page 16)*, installed as a floating floor *(page 21)*— boards glued to each other, but not fastened to the subfloor—or glued to the subfloor *(page 23)*. Not all methods suit all products, so ask about installation options when you buy the flooring.

A Wood Sandwich
Engineered wood flooring is three or five layers of wood glued together. The top layer is an attractive piece of solid wood. Thick top layers wear longer. The top layer may be rotary cut (peeled from the outside of the log) or sliced (cut from the middle like regular lumber). Rotary cutting gets more veneer from the same tree. Slicing creates a grain that looks more like solid wood.

STRIPS

Engineered strips come in a range of widths, starting at 2 1/2 inches. Most boards are prefinished. Exotic woods may be used for the top layer. These are cheaper than the same wood in solid flooring because only a thin layer is needed.

PLANKS

An engineered plank may be a single wide unit or be constructed of two or three strips laminated together into a wide board. The second option results in a look similar to a strip floor, but installs more quickly.

PARQUET FLOORING

A parquet floor looks like a jigsaw puzzle, but it's easier to do. The individual pieces of wood that make up the pattern are usually held together as larger tiles. Common sizes include 9x9, 11x11, 12x12, and 19x19.

Parquet tiles can be solid wood or engineered, as well as prefinished or unfinished. They're usually thinner than strip or plank flooring, but are just as durable.

Square tiles available at home centers are usually tongue and groove and are fairly easy to install. The more complicated patterns you can buy from a flooring dealer are more likely to be square edged. You might want to have these installed professionally.

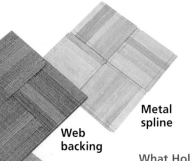

Metal spline

Web backing

Paper facing

What Holds Them Together?
Parquet tiles are held together by a paper facing that you peel off after installation, a web backing, or a metal spline. The spline is soft enough to cut with a power saw, so you can replace a tile if you need to.

PARQUET PATTERNS

It's not always easy to see how the pattern on the individual tiles will look when it's a larger pattern on your floor. And it's even harder to know if you'll like it. Flooring dealers may have photos or installed sections to give you a better idea.

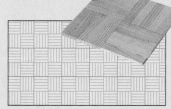

Finger block

Straight line (end-to-end)

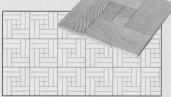

Monaco

Double herringbone

Decorative Insets
Dress up a wood floor with a parquet inset or border. A border can be both decorative and functional. Inlaid around the edges of the dining area or in an open-plan living area, it is a source of visual interest in itself and defines the space.

CHOOSING WOOD FLOORING

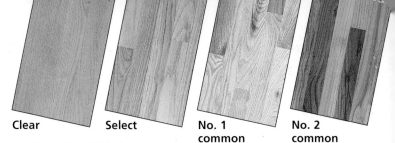

| Clear | Select | No. 1 common | No. 2 common |

Your choice will be partly about look, partly about properties of the wood, and partly about cost. In this chart, common hardwoods are grouped by the way they look: ash is a coarse-grained wood similar to the oaks; maple, birch, and beech are all lighter, finer-grained woods. A few exotic woods, good choices for insets or borders, are on page 6 with the softwoods.

The wear rating tells you how well a wood resists ➤

Getting a Good Grade
Wood is sorted and sold in different grades. The grade names and criteria depend on the species. Oak flooring grades are shown above. For many species, only the highest grade is usually sold for flooring, but for oak, and sometimes for other species, other grades are available. There's a big difference in appearance between the highest and the lowest grades. There's also a big difference in price. Lower grades can make fine flooring, but be sure to compare the same grade when you're price shopping.

WOOD SPECIES FOR FLOORING

COMMON HARDWOODS

SPECIES	Oak, red	Oak, white	Ash	Maple	Birch	Beech
WEAR	Very good	Excellent	Very good	Very good	Very good	Very good
STABILITY	Good	Good	Good	Poor	Poor	Poor
RELATIVE COST	Moderate	Moderate	Moderate	Moderate	Moderate	Moderate
AVAILABILITY	Usually a stock item	Usually a stock item	May be a special order	Usually a stock item	May be a special order	May be a special order

dents and stands up to heavy foot traffic.

The stability rating refers to how much the wood will expand and contract due to humidity. The less movement, the better.

Actual costs vary across the country, but relative costs are fairly stable.

Availability tells you if you'll have to place a special order for the wood you like.

The characteristics of the wood are not as important for engineered wood flooring. The layers provide stability, so you can make a choice based more on look and cost than on structural properties of the wood.

Wood Trim
When you're choosing the wood for your floor, it's a good time to think about installing new wood trim, too. Reducer strips, end caps, shoe molding, stair nosing—all are available in a range of woods to match your wood floor, whether it's solid wood or engineered.

WOOD SPECIES FOR FLOORING

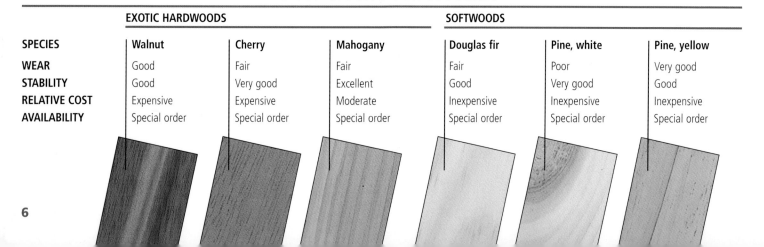

SPECIES	EXOTIC HARDWOODS			SOFTWOODS		
	Walnut	Cherry	Mahogany	Douglas fir	Pine, white	Pine, yellow
WEAR	Good	Fair	Fair	Fair	Poor	Very good
STABILITY	Good	Very good	Excellent	Good	Very good	Good
RELATIVE COST	Expensive	Expensive	Moderate	Inexpensive	Inexpensive	Inexpensive
AVAILABILITY	Special order	Special order	Special order	Special order	Special order	Special order

CALCULATING FLOORING NEEDS

To know how much flooring to buy, you need to know the room's square footage. Multiply length times width. Divide irregularly shaped rooms into sections and calculate the area of each. Work in inches instead of feet, so you don't have to deal with twelfths every time you measure something a few inches more than the nearest foot. Add 10 percent for waste and errors. Keep the leftovers for future repairs.

Strip and plank flooring is sold in bundles of different sizes. Take your room's measurements to your dealer.

Area coverage for parquet tiles is usually written on the package. If it isn't, multiply the area of each tile by the number of tiles.

Calculating Square Footage

Calculate the area in square inches for each part of the room. Add 10 percent for waste and error. You want your answer in square feet; divide by 144 to get there. In your calculations, round up numbers to the next whole number. Add the totals for each area together.

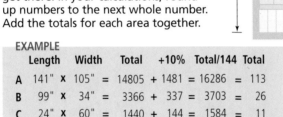

EXAMPLE

	Length		Width	Total		+10%	Total/144		Total
A	141"	x	105"	= 14805	+	1481	= 16286	=	113
B	99"	x	34"	= 3366	+	337	= 3703	=	26
C	24"	x	60"	= 1440	+	144	= 1584	=	11
							Total square footage		150

YOUR ROOM

	Length		Width	Total		+10%	Total/144		Total
A	___	x	___	= ___	+	___	= ___	=	___
B	___	x	___	= ___	+	___	= ___	=	___
C	___	x	___	= ___	+	___	= ___	=	___
D	___	x	___	= ___	+	___	= ___	=	___
							Total square footage		___

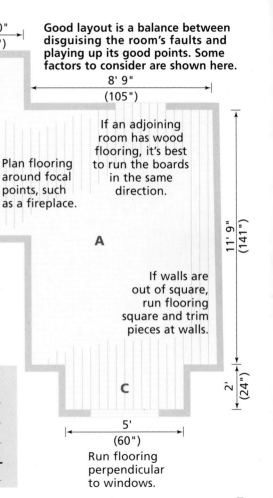

Good layout is a balance between disguising the room's faults and playing up its good points. Some factors to consider are shown here.

2' 10" (34")

8' 9" (105")

B

8' 3" (99")

If an adjoining room has wood flooring, it's best to run the boards in the same direction.

Plan flooring around focal points, such as a fireplace.

A

11' 9" (141")

If walls are out of square, run flooring square and trim pieces at walls.

2' (24")

C

5' (60")

Run flooring perpendicular to windows.

TOOLS FOR WOOD FLOORING

If you already have a well-stocked tool box, you probably won't need to buy many new tools to install wood flooring. You might want to rent a few, though.

The tools you need will depend on the type of flooring you're installing, so look at the installation instructions before you rush out and buy new stuff.

MEASURING & MARKING TOOLS

1. Level
Checks subfloor for level.

2. Chalk line
Marks working lines on floor.

3. Framing square
Keeps cutting lines straight.

5. Tape measure
For accurate measurements.

4. Compass and china marker
Transfers unevenness of wall to boards to lay out cut.

CUTTING & DRILLING TOOLS

1. Power miter saw
Makes cutting boards on an angle fast and easy.

2. Circular saw
For removing old floor boards and cutting new boards.

3. Saber saw
For cutting engineered boards to fit at edge of floor.

9. Jamb saw
Cuts door jamb and trim to fit flooring.

8. Butt chisel
To pry baseboards from wall and remove old flooring.

10. Miter box and backsaw
Cuts boards at precise angles.

7. Utility knife
For cutting sheets of foam underlayment.

4. Hammer drill
Drills holes in concrete.

5. Variable-speed electric drill
Drills pilot holes and sets screws.

6. Bits
Powerbore, Forstner, and combination bits drill holes for plugs or to remove damaged flooring; masonry bit drills holes in concrete.

INSTALLATION TOOLS

1. Power nailer
Shoots nails into board tongues.

2. Hand plane
To trim or smooth floor boards to fit.

3. Caulking gun
For applying construction adhesive.

4. Ball-peen hammer
Heavy, large-faced hammer for hitting chisels. Could substitute claw hammer.

5. Rubber mallet
For tapping boards in place.

6. Claw hammer
All-purpose hammer.

7. Nail set
For setting nails below surface of floor.

8. V-notched trowel
For spreading tile adhesive.

9. Putty knife
Transfers tile adhesive to subfloor.

15. Belt sander
To smooth high spots on wood subfloors or edges of floor boards to make them fit.

14. Pry bar
For prying off trim.

13. Crow bar
Pries balky boards into place.

12. Alignment bar
For positioning final boards at walls when installing floating floor. Manufacturer specific. Could substitute pry bar.

11. Tapping block
Specially shaped for tapping engineered boards.

10. Trowel
For spreading leveling compound.

FINISHING & REFINISHING TOOLS

1. Synthetic lamb's wool applicator
For spreading oil-base finish. Rinse twice in clean paint thinner and dry before use.

2. Foam applicator
For spreading water-base finish.

3. Wide paintbrush
For applying finish along the wall, in corners, and in other tight areas. Use synthetic brush or painter's pad for water-base finish.

6. Edge sander
For sanding in tight corners and against walls.

5. Pad sander
Does the major sanding of the floor.

4. Sanding block
For sanding pieces by hand.

SAFETY GEAR

Wear the appropriate safety gear. Make sure you have:
- A respirator when working with toxic materials.
- A dust mask when sanding.
- Safety goggles to prevent scraps and chips from damaging your eyes.
- Rubber gloves when working with caustic materials.
- Ear plugs. Loud power tools will damage your hearing.
- Knee pads if you'll be kneeling frequently.

UNDERLAYMENTS, ADHESIVES, AND FINISHES

For a good-looking, long-lasting floor, start by choosing the right product for your situation. Some wood flooring is not suitable for installation on or below grade. Check requirements with your dealer.

Depending on the flooring and where you're installing it, you might need a plywood underlayment, a vapor barrier, or a plastic-foam underlayment. If you're fastening the flooring to the subfloor, you'll need adhesive—except for self-stick parquet tiles.

For unfinished flooring, you'll need a finish as soon as the floor is laid. For prefinished boards, you'll need one when the existing finish wears off.

WHAT ELSE GOES UNDERNEATH?

What goes underneath depends on the kind of flooring and where you're installing it.

◆ Polyethylene (plastic) vapor barriers are used over concrete subfloors to keep moisture from creeping into the wood. They're not a solution to a moisture problem, just insurance in normal situations. In some areas, vapor barriers are also advised over wood subfloors. Ask your dealer.

◆ Rolled plastic-foam underlayment acts as a cushion and a sound deadener under floating floors. The material is laid loose on the subfloor or existing floor.

WOOD UNDERLAYMENTS

Sometimes you can lay new wood flooring over an existing subfloor. An extra layer of plywood underlayment adds strength and provides a smooth surface. Smoothness is especially important under parquet. For this, 1/2-inch or 3/8-inch underlayment-rated plywood will do fine.

To lay strip or plank flooring parallel to the joists, you need an extra-strong subfloor—a minimum of 1 1/8 inches is recommended. Adding a layer of 5/8-inch plywood over an existing subfloor does the trick.

DEALING WITH DAMP CONCRETE

If your concrete fails the moisture test *(page 12)*, fix the problem before you install a wood floor. A plastic vapor barrier isn't enough—you need a professionally installed moisture barrier. One type is shown here, but other kinds are available. They all reduce moisture vapor emissions from concrete so that finish flooring can safely be installed. Specify to the professional you contact that you want to be able to install a wood floor.

Finish coat

Base coat

Fiber mat

ADHESIVES

It's easy to get the right adhesive for the flooring you're installing—just read the label. It'll tell you:
- What type of flooring it holds down (strip or parquet)
- What wood species it's good for
- How many square feet the can covers
- What size notched trowel to buy
- What product to use to clean up

Pay attention to the adhesive's "open time," which is also marked on the label. This is the length of time the adhesive is workable before it dries. Only apply as much adhesive as you can cover with flooring before it dries. If it dries before you've set your flooring, scrape it off and apply new.

WISDOM OF THE AISLES

If you find dried adhesive smears on the flooring surface after you've finished your installation, don't panic. They'll clean up easily with a citrus-based cleaner. Just use a clean rag and follow the product directions on the container.

FINISHES

Penetrating sealers soak into the wood and harden, sealing it against dirt and some stains. They're available combined with a stain for added color. A wax finish applied on top adds abrasion resistance and a soft shine. Don't use sealers and wax where they might get wet.

Surface finishes, on the other hand, stay on top of the floor, forming a durable, moisture-resistant surface that's fine for use in kitchens and other potentially wet areas. They shouldn't be waxed.

SURFACE FINISHES	CHARACTERISTICS	APPLICATION
Polyurethane (oil-modified urethane)	Durable; lasts longer than water-based finish; moisture-resistant; tends to yellow with age; available in gloss, semigloss, satin, or mat	Easy to apply; can apply 1 coat every 4 hours
Water-base finish	Clear; durable; harder than polyurethane finish; fast drying; nonyellowing; available in gloss, semigloss, satin, or mat	Easy to apply; can apply more coats in less time than polyurethane
Moisture-cured urethane	Hardest and most moisture-resistant; nonyellowing; gloss only	Difficult; get a professional
Swedish finish	Clear; very durable; generally harder than polyurethanes; fast drying; resists yellowing	Difficult; get a professional

PREPARING THE AREA

LAST-MINUTE CHECKLIST

Make sure you have:

✓ Jamb saw
✓ Pry bar
✓ Utility knife
✓ Plastic sheeting
✓ Duct tape

Whichever type of wood flooring you are laying down, start the project by removing the baseboards. Work carefully so you can replace the same boards after you've installed the floor.

If there's door molding in the way, you don't have to remove it. Instead, cut enough from the bottom to make room for the new floor.

Wood flooring can only be installed over a dry subfloor. If you plan to lay your new floor on concrete, check the surface for moisture since concrete floors are prone to dampness. If you detect problems, refer to page 10 for remedies.

Removing the baseboards
If there's paint bonding the top edge of the baseboard to the wall, cut through the paint with a utility knife.

Gently wedge the baseboard away from the wall with a pry bar, protecting the wall with a wood scrap.

WISDOM OF THE AISLES

One of the simplest ways to deal with door molding is to leave it in place and cut a space at the bottom with a jamb saw. Use a piece of flooring, and underlayment if it's being installed, as a gauge of how much to trim off the molding. We'll show you how on page 23.

Checking a concrete floor for moisture
Tape a piece of clear plastic sheeting to the floor. Pull the plastic up after 24 hours—if there's moisture on it or the floor is discolored, you have a problem. You'll need professional help to fix it. See page 10.

SMOOTHING THE FLOOR

LAST-MINUTE CHECKLIST

Make sure you have:

✓ Level
✓ Straight 1x4
✓ Hammer
✓ Nail set
✓ Belt sander
✓ Putty knife
✓ Leveling compound
✓ Trowel
✓ Cold chisel
✓ Ball-peen hammer

Besides being dry, the subfloor for wood flooring needs to be level and flat. An uneven surface can lead to all kinds of problems, such as creaking and loose floorboards. You can live with a floor that has a gentle slope, however. Up to $\frac{1}{2}$ vertical inch across 10 feet of floor is acceptable. If the slope is greater than this, you'll have to level the subfloor. And you'll probably need help doing it. Call a professional.

As shown here, minor high and low spots on an otherwise level floor are easy to fix. Sand or chip off the high spots and fill in the low spots.

1

Checking for level and flat
Place a 1x4 on edge on the floor and set a level on it. Low spots will show up as gaps under the board. Check the level to see if the floor slopes. Test a few places.

2

Flattening a wood floor
Flatten small high spots with a belt sander. Sink protruding nails with a hammer and nail set first so they don't rip the sandpaper. Fill low spots with a wood leveling compound.

3

Smoothing a concrete floor
Level high spots by chipping them down with a ball-peen hammer and a cold chisel. Fill low spots with a leveling compound, applying it with a putty knife and spreading it with a trowel.

13

LAYING PLYWOOD ON CONCRETE

Laminated planks and glue-down parquet can be laid directly on concrete, but other types of wood flooring generally need plywood underlayment. Support the plywood with a grid of 2x2s. Glue them down with construction adhesive, then fasten them with concrete anchors. Drill holes for the anchors with a hammer drill and a diamond-tipped bit. Let the bit do the work; don't push. Cut foam insulation $1\frac{1}{2}$ inches thick to fit in the spaces in the grid. It adds support under the plywood and insulates the floor. Insulation might not be needed in all climates; ask your dealer. Finally, glue the plywood subfloor to the foam and nail it to the 2x2s.

1 Starting the grid
Cover the floor completely with strips of plastic sheeting. Place 2x2s on the sheeting along the long walls and 16 inches on center in between. Run a bead of construction adhesive along the top of each 2x2. Turn the 2x2s over to bond them in place.

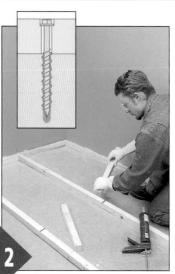

2 Finishing the grid
Cut short pieces of 2x2 and glue them down at 48-inch intervals to complete the grid. Drill holes through the 2x2s and into the concrete at 12-inch intervals. Insert concrete anchors (inset).

3 Laying the subfloor
On a table saw, cut sheets of 1½-inch-thick foam insulation to fit and place them into the grid. Spread adhesive on the sheets, then cover them with panels of underlayment-rated plywood. Nail the panels every 3 inches around the perimeter and 6 inches on the 2x2s in the center.

LAYING PLYWOOD UNDERLAYMENT

Wood flooring can be laid right over an existing resilient sheet or tile floor, provided any loose flooring is first glued down. If the existing floor is wood or ceramic tile, however, you'll need to pull it up. In the case of ceramic tiles, you'll have to lay down plywood underlayment on top of the old subfloor, as shown here. Nail down CD-grade plywood, ⅝ inches thick, making sure the ends and edges of the panels are offset from those of the old subfloor. Don't butt the panels too tightly together; leave a ⅛-inch gap for expansion and contraction. You can use nails, as shown here, or screws. Don't use drywall screws; they'll snap off.

1

Arranging the panels
Lay the plywood sheets on the old subfloor, offsetting ends and edges of the two layers by at least 3 inches. Leave a ⅛-inch gap between panels.

2

Fastening the underlayment
Anchor the underlayment by driving nails 3 inches apart around the perimeter of the panels. Nail the center of each sheet in rows 6 inches apart.

ACCLIMATE THE FLOORING

Once you've prepared the room, smoothed the subfloor, or added new plywood, you may need to wait a few days before you install the flooring so the wood has time to acclimate to its new environment. When you bring the flooring home, let it adjust to the humidity level in your home before you install it.

Rip open the packages, but leave the binding on. Let them sit about a foot from the wall, to allow for air circulation, for about seven days. How long the flooring needs to acclimate depends on your climate; ask your dealer.

INSTALLING STRIP OR PLANK FLOORING

All tongue-and-groove flooring, whether strips or planks, is installed in pretty much the same way. Each board is fastened to the subfloor with nails or staples driven through the tongue. The fasteners are concealed by the next board installed. The only exceptions are the first and last boards along the starting and finishing walls. Both are face-nailed to the subfloor—nails are driven straight through them along the outside edges. See page 18 for the right size of nail or staple to use.

Planks wider than about 4 inches must be blind-nailed and then screwed down through the face *(page 20)* to minimize warping.

1 Aligning the first row
Position the longest board parallel to the starting wall so its groove edge is $1/2$ inch away. Mark the edge of the board under the tongue on the floor at each end. Snap a chalk line between the marks.

2 Arranging the boards
Open several packages of boards and lay out a dry run. You'll want to shuffle the boards to achieve the best grain and color mix. Place strips so the ends in successive rows are offset by a few inches.

3 Face-nailing the first row
Place the first board along the starting wall, aligning the edge under the tongue with the chalk line. Drill pilot holes through the board every 6 inches, $1/2$ inch from the groove edge. Drive the nails until the heads are just above the board, then sink them with a nail set.

4

Drilling for blind nails

Put a drill bit the diameter of the nail in your drill. Set the bit in the corner formed by the tongue and edge of the board. Angle it about 45 degrees and drill pilot holes every 6 inches through the strip.

5

Blind-nailing

Drive a nail partway into each pilot hole, then sink the head below the surface with a nail set. Install the remaining boards of the first row the same way, fitting the end tongues and grooves together snugly.

6

Framing around an obstacle

Miter boards to fit around the obstacle, so the grooves face the appropriate direction for the rest of the floor. Drill pilot holes and drive nails partway in on the groove side of boards. Apply carpenter's glue to mitered ends, fit the boards, and seat nails. Drive nails through tongues about every 6 inches. Fill between the boards and obstruction with silicone caulk.

7

Laying subsequent rows

Install each board of the remaining rows so the tongue of the previous row fits in the groove of the new row. To seat the tongue in the groove snugly, fit a scrap piece of flooring against the outside board and tap it with a rubber mallet. Blind-nail the boards, as described in Step 5.

1 7

INSTALLING CROOKED BOARDS

To straighten a bowed strip or plank while you blind-nail it, temporarily screw a board to the subfloor a few inches in front of the curve, and use a crow bar and a scrap piece of flooring to force the board into alignment. If the floor board is also cupped, you can flatten it at the same time with your foot and a two-piece jig made from a 2x4 and a triangular wood block.

8

Using a power nailer

Once you've fastened five or six rows, you'll be far enough from the starting wall to use a power nailer or stapler. The tool speeds up installation— it drives the nails quickly without the need for pilot holes. Strike as often as needed to drive the fastener.

FASTENING TONGUE-AND-GROOVE FLOORING

FLOORING THICKNESS	LENGTH OF FASTENERS	FASTENER SPACING
3/4"	2" with power nailer; 2 1/4" or 2 1/2" by hand	8" (plank); 10" (strip)
1/2"	1 1/2" or 1 3/4" with power nailer; 1 3/4" or 2" by hand	10"
3/8"	1 1/4" with power nailer; 1 1/2" or 1 3/4" by hand	8"

9

Adjusting the last rows

The last row ends about 1/2 inch from the wall. You can shave up to 1/8 inch off the last board to make it fit. If you have more than 1/8 inch to take off, shave boards in as many rows as necessary. Draw a guideline and put the board groove-edge up in a vise. Make minor adjustments with a hand plane and larger ones with a belt sander.

10

Nailing the last row

Drive nails partway into flooring near the tongue. Wedge boards in place with a board pressed against two boards placed flat against the wall. Drive the nails.

11

Sanding the reducer strip

A special piece of flooring, called a reducer strip, eases the transition between the new wood floor and the existing floor in an adjoining room. Sand it with 80- or 120-grit sandpaper in a sanding block.

12

Installing the reducer strip

Set the reducer strip in place so its groove fits over the tongue of the last board and its rounded edge laps over the flooring in the adjoining room. Face-nail the strip in place.

13

Reinstalling the trim

For unfinished boards, apply a finish (pages 26-28) before you reinstall the trim. For prefinished boards, you can reinstall the trim now. Place each piece against the wall you removed it from. Fasten the baseboards to each wall stud with $1\frac{3}{4}$-inch finishing nails. Fasten the shoe molding to the baseboards—not to the floor—with $1\frac{1}{4}$-inch finishing nails.

SCREWING DOWN PLANK FLOORING

If your floorboards are less than 4 inches wide, they won't need to be screwed down. With wider planks, however, you'll need to screw through the face of the boards so that they don't warp. Nail down the entire floor of boards first, then install the screws. Set the heads of the screws below the surface of the planks and fill the holes with wood plugs.

Dealers who sell wide planks also usually offer plugs in a range of species and sizes. Create a decorative effect with plugs in a different wood from the floor, as shown here, or choose plugs of the same wood as the floor to blend in.

1. Marking screw holes
Drill holes for marking plug holes in the end of a wood block. Where you place the holes depends on the size of the boards and plugs. Test on a scrap of flooring until you get the look you want. Align the edge of the block with the end of a plank and mark the holes. Mark each plank.

2. Drilling and anchoring
With a Forstner or spade bit, drill a hole at each mark as deep as your wood plugs are long. Choose the right size bit for the size of plugs you're using. Then drive a 1½-inch No. 8 wood screw into each hole *(inset)* so the head is seated on the bottom.

3. Plugging the screw holes
Fit a wood plug into each hole, then tap the plugs flush with the floor, protecting the surface with a wood block. Any slight unevenness will be sanded off when you sand the floor prior to finishing.

LAYING A WOOD FLOATING FLOOR

LAST-MINUTE CHECKLIST

Make sure you have:

✓ Flooring
✓ Foam underlay
✓ Glue
✓ Compass
✓ Saber saw
✓ Framing square
✓ Alignment bar
✓ Hammer
✓ Utility knife
✓ Pry bar

Engineered wood flooring often can be installed as a floating floor right over existing flooring. The boards are glued together tongue to groove, and laid over strips of plastic-foam underlayment. The flooring is not fastened to the subfloor, but is held down by its own weight. A vapor barrier always goes down first over concrete and, in some areas, over wood; ask your dealer.

If your flooring comes with an alignment bar to fit boards in tight places, use it. If not, use a pry bar. Follow the manufacturer's directions for the minimum size of cut boards at the ends of rows.

Use the manufacturer's glue. If you don't, you might void your warranty.

1

Laying down the underlayment
Unroll strips of plastic-foam underlayment, cutting ends square so you cover the old floor completely. Butt the strips together edge to edge.

2

Aligning the first row
You'll need to trim the first row if the starting wall isn't straight. Dry-lay the boards straight, with a spacer to keep the groove edge ½ inch from the wall. Transfer the wall line to the boards with a compass. Trim the boards along the line with a saber saw.

3

Setting the first board
Position the first board with its groove edge ½ inch from the wall and, lifting it slightly, spread a bead of glue in the end groove.

4

Finishing the first row
Fit the tongue of the second board into the end groove of the first. Close the joint as in Step 5. Wipe up any excess glue. At the end wall, snug up the joint with an alignment bar. Fit its flange on the end at the wall and tap the bar to tighten. If your manufacturer doesn't supply an alignment bar, use a pry bar as in Step 7.

5

Aligning subsequent rows
Spread glue in the grooves of the first board of the next row. Then fit the groove over the tongue of the first row. Close the joint with a hammer and tapping block or a scrap of wood placed against the edge of the board. Lay down subsequent rows the same way until you run out of space.

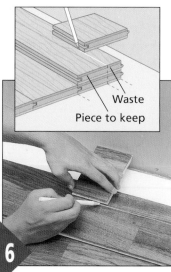

Waste

Piece to keep

6

Trimming the last boards
To cut the last row to fit at the wall, place a board on the second-to-last row, aligning the edges. Place a scrap board on top of the stack so one edge butts against the wall, then mark a line on the top board. Measure and mark a new line ½ inch inside the first line, to leave expansion room at the wall. Cut board along new line.

7

Fitting the last row
Close the tongue-and-groove joints between the boards of the last two rows with an alignment bar or a pry bar. Fit a wood pad between the bar and the wall, so as not to damage the wall. Reinstall any trim you removed, as described on page 19.

INSTALLING A PARQUET FLOOR

LAST-MINUTE CHECKLIST

Make sure you have:
- ✓ Parquet tiles
- ✓ Adhesive
- ✓ Framing square
- ✓ Chalk line
- ✓ Trowel
- ✓ Putty knife
- ✓ Notched trowel
- ✓ Rubber mallet
- ✓ Table saw

Like strip and plank flooring, parquet tiles usually have tongues and grooves, but that's where the similarity ends. Parquet is glued down, rather than nailed. Installation starts at the center of the room so that if tiles have to be trimmed to fit at the walls—and they usually do—all the outside tiles can be cut by the same amount. Laying a dry run, as shown here, will show you how much.

Trim tiles on a table saw fitted with a carbide precision-cut blade. Center 2-inch masking tape over the cut line to prevent tear out.

Open the packages and let the tiles sit in the room to acclimate for 48 hours before installation.

1

Preparing the room

Lay a tile on the subfloor, butt it up against the door molding, and cut the trim just above the tile with a jamb saw. This ensures you'll be able to lay tiles right up to the wall.

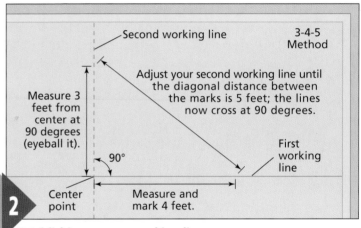

Second working line

3-4-5 Method

Adjust your second working line until the diagonal distance between the marks is 5 feet; the lines now cross at 90 degrees.

Measure 3 feet from center at 90 degrees (eyeball it).

First working line

90°

Center point

Measure and mark 4 feet.

2

Establishing square working lines

Mark the centers of two opposing walls and snap a chalk line on the subfloor between the marks. This is your first working line. Mark the center point of this line. Measure and mark a point on this line 4 feet from the center. To position your second working line, mark the centers of the opposite walls and string a chalk line between these two marks. Applying the 3-4-5 triangle principle shown above, check that the chalk line crosses the first working line at 90 degrees. Snap the chalk line when it's at 90 degrees.

3

Laying a dry run

Lay a row of tiles from the center point to one wall, fitting tongues and grooves together. If the last tile needs to be cut by more than half, move the line back by the width of half a tile. This will create balanced cuts on opposite walls.

4

Laying out in a hallway

Snap a chalk line on the subfloor between the midpoints of the two end walls. Make a dry run from the line to one side wall—reposition the line to avoid having to cut the last tile by more than half its width, as explained in Step 3.

5

Spreading the adhesive

With a putty knife, lay enough adhesive for one to three tiles at the starting point. Spread it in a thin layer with an 8-inch V-notched trowel held at a 45-degree angle.

6

Laying the first tile

Clean out the grooves of each tile before you set it. Set the tile on the adhesive so one corner aligns with the intersecting chalk lines. Press the tile into the adhesive.

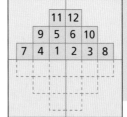

INSTALLATION SEQUENCE

Lay tiles in the pyramid pattern shown here. Tiles can vary by as much as 3/32 inch, making them hard to align. Laying them like this helps because you fit most tiles into a corner made by neighboring tiles. They still won't align perfectly, but the variations won't throw you off.

7
Covering the floor
Continue laying tiles, fitting the tongues of one tile into the grooves of the next. Level the tiles by tapping them into the adhesive with a rubber mallet.

8
Laying tiles at a wall
If you need to cut tiles to fit, place a tile on the last one installed, aligning their edges. Place a third tile, the "marker tile," on top of the stack so one edge butts against a $1/2$-inch spacer placed against the wall. Mark a line along the other edge. Cut the tile along the line. Make sure that the hot-melt tape that helps hold the tile together won't interfere with the joint. Peel it off if necessary.

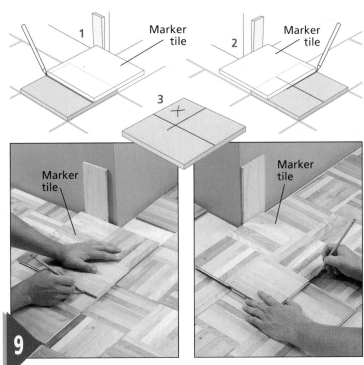

9
Cutting a tile to fit at a corner
Mark a cutting line on the tile to fit along one of the two walls (left), as described in Step 8. Then move the tile to be cut to the adjoining wall—without turning the tile, so the line drawn on it is parallel to the same wall. Butt the marker tile against the wall and trace along it, making a line that intersects the other marked line (right). Label the waste section formed by the two lines with an X, then cut the tile.

LAST-MINUTE CHECKLIST

Make sure you have:
- ✓ Pad sander
- ✓ Edge sander
- ✓ Sandpaper
- ✓ Putty knife
- ✓ Wood paste
- ✓ Paintbrush
- ✓ Lamb's wool applicator
- ✓ Wood finish

If you're installing a new floor, you're at the beginning of the end. If you're refinishing an old floor, you're at the beginning of the beginning. Either way, you're in for some work.

The process is the same, whether you're finishing a new floor or refinishing an old one. To refinish engineered wood, though, it's best to hire a professional. Even with a pad sander, you might sand through the top layer.

A pad sander, *as shown here*, always requires four sandings: a first pass with medium-grit paper to rough up the floor, then one pass each with coarse-, medium-, and fine-grit papers. You can get the papers where you rent the sander.

Setting up the pad sander
Place a square of sandpaper on the floor with the rough side facing down. Put the white pad on top of the paper to cushion the machine. Next, angle the sander over the pad and sandpaper and lower it down.

Ask the salesperson where you rent the machine how to load the sandpaper.

CHECK FOR WAX

Test for paste wax before sanding an old floor. If wax is left on the floor, it will clog the sandpaper. Test by wiping a white rag dampened with turpentine over the floor. Bright orange or yellow on the rag means there is paste wax on the floor. Remove it with turpentine or a wax remover before you sand.

Sanding the floor
Wear goggles, ear muffs, and a respirator. Start at one wall and guide the sander to the opposite wall, moving parallel to the flooring grain. Move the sander to one side and sand a new strip, overlapping the first by a few inches. Once the the whole floor is done, make a pass with coarse sandpaper.

3

Filling cracks

Fill any cracks with a liquid wood paste. Pour the paste onto the floor and push it into the cracks with a putty knife. Allow 24 hours to dry, then sand twice more—once with medium-grit paper and once with fine-grit.

4

Brushing on the finish

Prepare the finish following the manufacturer's directions, then apply it along walls and to other tight spots with a wide paintbrush. Use a synthetic-bristle brush or a disposable foam paintbrush for water-base finishes.

5

Spreading finish on the rest of the floor

Cover the main part of the floor with back-and-forth passes that overlap slightly, following the grain. Use a synthetic lamb's wool applicator (above) for oil-base finishes, a foam applicator for water-base finishes. Let the finish dry and apply more coats according to the product instructions.

REFINISHING TIPS

◆ First empty the room, taking out all the furniture and window coverings. Then remove the shoe moldings.

◆ A painted floor needs an extra pass with coarse sandpaper.

◆ Buy at least a dozen sheets of sandpaper in each grit category for both the pad sander and edge sander. Most suppliers will credit you for unused paper when you return the sanders.

◆ The dust produced by sanding a wood floor is highly flammable. Seal the room by taping plastic sheeting around doorways and ducts; make sure the room is well ventilated. Set up an exhaust fan in a window to increase air circulation; tape a furnace filter to the room-side of the fan to clean dust particles from the air.

◆ Do not smoke when applying a finish.

◆ Wear a respirator, and keep windows and doors in the room open when applying a finish.

STAINING A WOOD FLOOR

You might love the color of your floor the way it is, but if you don't—stain it.

◆ Oil-base stains are the traditional choice and still very popular. Water-base stains are less toxic, but tend to raise the wood grain, making the surface rough. Fast-dry stains are like oils, but with additives to make them dry faster.

◆ Get a stain and finish that are compatible. If the information is not on the labels, ask your dealer or call the manufacturer.

◆ Sanding will change the color of the wood a bit, so wait until the rough sanding is over to choose your stain color.

◆ Staining is a two-person job: one applies the stain and one wipes it off with a rag. Use lint-free, absorbent rags for the best results.

◆ Start in an inconspicuous area until you get the hang of it.

◆ To test if the stain is dry before finishing, rub a white rag on it. If the rag stays clean, the stain is dry.

SANDING THE EDGES

For most newly installed flooring, you won't need an edge sander. A pad sander can get close enough to the wall so the unsanded strip will be covered by the baseboard and shoe molding. Sometimes, especially in refinishing where you might not want to remove the baseboards, the pad sander won't get close enough. You'll need to sand the edges with another tool. An edge sander makes fast work of the job. Load it with sandpaper and arc it from side to side. For a small job, you could use a palm sander, but it'll take longer.

REPAIRING A WOOD FLOOR

LAST-MINUTE CHECKLIST

Make sure you have:

- ✓ Drill and bit
- ✓ Circular saw
- ✓ Butt chisel
- ✓ Utility knife
- ✓ Rubber mallet
- ✓ Glue
- ✓ Hammer
- ✓ Wood flooring
- ✓ Finishing nails

Wood flooring will usually forgive harsh treatment. Small cracks and gouges can be disguised with wood filler. Scratches and stains affecting only the surface can often be eliminated by sanding and refinishing. But if the damage runs deeper or several boards are affected, you may have no choice but to cut out the damaged boards and patch the area with new flooring. You'll need to make the cutting depth equal to the thickness of the flooring. Remove a baseboard to measure the thickness of the floor and to see which direction the tongues face.

Keep pieces you remove to make a custom filler for future repairs. Cut the pieces up for sawdust, mix it with glue, and fill dents.

1

Cutting the damaged boards
With a 5/8-inch bit, drill holes across and just inside each end of damaged boards *(left)*. Adjust the blade depth of your saw to the thickness of the flooring. Then make a series of cuts between each pair of holes at 1/4-inch intervals. Start in the middle and work toward the groove edge *(right)*. Avoid hitting nails on the tongue edge.

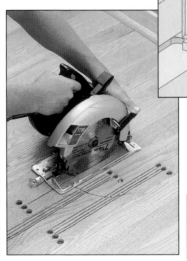

2

Removing the damage
Angle a 1 1/2-inch butt chisel into a saw cut and tap the handle to pry out the wood. The nails will come out with the wide piece. At the board ends, hold the chisel vertically, bevel facing the waste *(inset)*.

3 **Fitting replacement strips**
Cut floorboards to fit the opening. Trim the bottom lip from the groove of each piece. Apply finish *(pages 26-28)*. Apply wood glue to the tongue of the adjoining board. Angle the first board into place so its groove fits onto the tongue of the adjoining board. Seat the board by tapping a scrap piece of flooring against it with a rubber mallet.

4 **Fastening the boards**
Blind-nail each board to the subfloor as you would when installing a new floor *(page 17)*.

5 **Fitting the last board**
When you get to the last board, spread wood glue on the tongue of the second-to-last replacement board, then angle the last one into place.

6 **Pressing the board in place**
To level the replacement boards with the surrounding floor, place a board across the patch and tap it with a rubber mallet until all the boards are fully seated.

REPAIRING A PARQUET FLOOR

LAST-MINUTE CHECKLIST

Make sure you have:
- ✓ Circular saw
- ✓ Hammer
- ✓ Butt chisel
- ✓ Sandpaper (120-grit)
- ✓ V-notched trowel
- ✓ Parquet-tile adhesive
- ✓ Putty knife

If the time comes when you have to patch a damaged parquet floor you installed yourself, you'll appreciate having bought more tiles than you actually needed to lay the floor. Finding parquet tiles that blend seamlessly with an existing floor is almost impossible —especially since wood changes color as it ages.

The technique for replacing damaged tiles is similar to the steps for repairing strip flooring: Remove the damaged section, then cut and install replacement tiles to fit the opening.

1

Cutting out the damage
Adjust the blade depth of a circular saw to the flooring thickness, then cut just inside the edges of the damaged section—in this example, a section of a single tile.

2

Preparing the opening
Place a butt chisel on any remaining waste, with the bevel facing the opening. Strike with a hammer to chip out the waste. Scrape old adhesive off the subfloor.

3

Laying a replacement tile
Cut the bottom lips off the tile's grooves. Smooth the edges of the tile with 120-grit sandpaper. Spread adhesive on the subfloor with a V-notched trowel and lay the tile.

Wood Flooring 1-2-3
Project Director: Benjamin W. Allen
Editor: Jeff Day
Associate Art Director: Tom Wegner
Copy Chief: Catherine Hamrick
Copy Editor: Terri Fredrickson
Contributing Proofreader:
  Margaret Smith
Electronic Production Coordinator:
  Paula Forrest
Editorial Assistants: Karen Schirm,
  Kathleen Stevens
Production Director:
  Douglas M. Johnston
Production Manager: Pam Kvitne
Assistant Prepress Manager:
  Marjorie J. Schenkelberg

Cover photograph:
  Doug Hetherington Photography

Meredith® Books
Editor in Chief: James D. Blume
Design Director: Matt Strelecki
Managing Editor: Gregory H. Kayko
Director, Sales & Marketing, Retail:
  Michael A. Peterson
Director, Sales & Marketing,
  Special Markets: Rita McMullen
Director, Sales & Marketing,
  Home & Garden Center Channel:
  Ray Wolf
Director, Operations:
  George A. Susral

Vice President, General Manager:
  Jamie L. Martin

Meredith Publishing Group
President, Publishing Group:
  Christopher M. Little
Vice President, Consumer Marketing
  & Development: Hal Oringer

Meredith Corporation
Chairman and Chief Executive
  Officer: William T. Kerr
Chairman of the Executive
  Committee: E.T. Meredith III

The Home Depot
Senior Vice President of Marketing:
  Dick Hammill
Project Director: Barbara Koller

**Book Development Team
St. Remy Multimedia Inc.**
President: Pierre Léveillé
Vice President, Finance:
  Natalie Watanabe
Managing Editor: Carolyn Jackson
Managing Art Director:
  Diane Denoncourt
Production Manager:
  Michelle Turbide
Director, Business Development:
  Christopher Jackson

Editorial
Jennifer Ormston, Marc Cassini,
  Gerard Dee, Emma Roberts, Brian
  Parsons, Pierre Home-Douglas

**Art, Design,
  Illustration, & Studio**
Francine Lemieux, Michel Giguère,
  Robert Chartier, Normand
  Boudreault, Anne-Marie Lemay,
  Maryo Proulx, Jean-Guy Doiron

Production & Systems
Dominique Gagné, Edward Renaud,
  Jean Sirois, Martin Francoeur,
  Sara Grynspan

Special thanks to:
Maryse Doray, Lorraine Doré,
  Ryan Cavell, Robert Labelle,
  Karl Marcuse, Michael Wells

Consultants
Jon Arno, Stewart McLaughlin

Acknowledgments
Anderson Hardwood Floors,
  Clinton, SC
Barwood Hardwood Flooring,
  Montreal, PQ
Boen Hardwood Flooring Inc.,
  Martinsville, VA
Les Bois M&M Ltée, St. Mathieu, PQ
Bruce Hardwood Floors, Dallas, TX
Carlisle Restoration Lumber,
  Stoddard, NH
CR Parquet Floors, Anderson, CA
National Oak Flooring Manufacturers
  Association, Memphis, TN
Nordic American Corporation,
  Atlanta, GA
Quincaillerie Notre Dame de
  St. Henri, Inc., Montreal, PQ
Satin Finish Hardwood Flooring,
  Weston, ON
Sealflex Industries, Inc.,
  Costa Mesa, CA
Senco Products, Inc., Cincinnati, OH
Stanley Bostitch, East Greenwich, RI
World Floor Covering Association,
  Anaheim, CA

The editors of *Wood Flooring 1-2-3*
are dedicated to providing accurate,
helpful, do-it-yourself information. We
welcome your comments about
improving this book and ideas for
other books we might offer.

Contact us by any of these methods:

Leave a voice message at:
  800/678-2093

Write to:
  Meredith Books,
  Wood Flooring 1-2-3
  1716 Locust St.
  Des Moines, IA 50309

Send e-mail to: hi123@dsm.mdp.com